*For Christopher Mendenz Griffin*

## STERLING CHILDREN'S BOOKS
New York

An Imprint of Sterling Publishing Co., Inc.
1166 Avenue of the Americas
New York, NY 10036

STERLING CHILDREN'S BOOKS and the distinctive Sterling Children's Books
logo are registered trademarks of Sterling Publishing Co., Inc.

ISBN 978-1-4549-2992-5

Distributed in Canada by Sterling Publishing Co., Inc.
c/o Canadian Manda Group, 664 Annette Street
Toronto, Ontario M6S 2C8, Canada
Distributed in the United Kingdom by GMC Distribution Services
Castle Place, 166 High Street, Lewes, East Sussex BN7 1XU, England
Distributed in Australia by NewSouth Books
University of New South Wales, Sydney, NSW 2052, Australia

For information about custom editions, special sales, and premium and corporate purchases,
please contact Sterling Special Sales at 800-805-5489 or specialsales@sterlingpublishing.com.

Manufactured in China

Lot #:
2   4   6   8   10   9   7   5   3   1
12/18

sterlingpublishing.com

**PHOTOGRAPHS:**
**COVER Front:** Seapics; **Back:** Cultura Creative
**INTERIOR Alamy:** Andrey Armyagov 28, ArteSub 14, Sabena Jane Blackbird 12, Mark Conlin  20, **ImageBroker:** 18,
Cultura Creative 16, © Jeff Rotman 19, Steve Trewhella 8 (catshark), Michael Wheatley 7, WaterFrame 23;
**Animals Animals:** © Bob Cranston 21, © Sea Images, Inc. 29; **ArdeaL** © Valerie & Ron Taylor: 6, 22;
**Getty images:** © Federica Grassi 17, © Mohamad Haghani/Stocktrek Images 9; **iStock:** Howard Chen 24, pjohnson1 29;
**Minden Pictures:** © Albert Llea 8 (Lantern), Jeff Rotman 8 (whitetip reef 2x); **Nature Picture Library:** © Niall Benvie 15,
© Doug Perrine 25 (sawfish); © **SeaPics** 13, 25 (goblin); **Shutterstock:** Andrea Izzotti 26, Tunatura 10;
**Stockfood:** © PhotoCuisine/Bono 27 (dogfish); **Superstock:** © Steve Bloom Images 11,
© Luca Tettoni/Robertharding 27 (soup), © Norbert Wu 25 (carpet)

## Just Ask!

# DO SHARKS GLOW IN THE DARK?

### . . . and Other Shark-tastic Questions

**MARY KAY CARSON**

### Are sharks fish?

**YES!** Sharks are ocean fish that are **carnivores**. That means they eat meat. Earth's seas are home to hundreds of kinds of sharks.

Shark bodies are built for underwater hunting. Sharp eyes spot **prey**. Powerful tails turn on the speed. Pointy teeth munch down. What **predator** power!

## Do they have bones?

# NOPE. The skeleton of a shark isn't bone. It's made of tough, bendable **cartilage**. That's the same stuff that gives your nose and ears their shape.

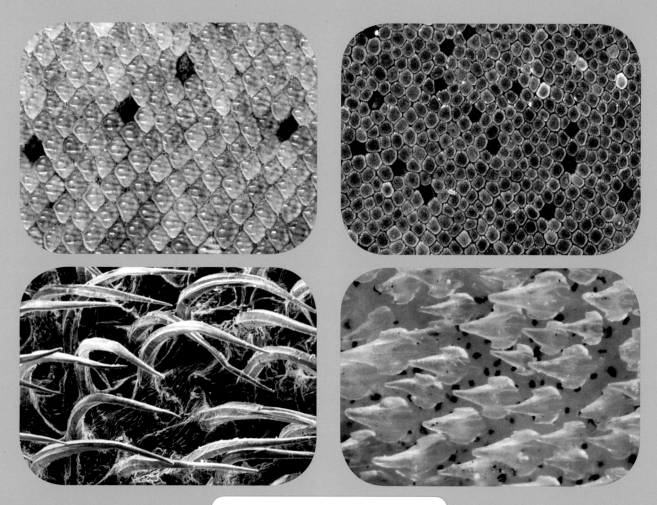

Look at the different types of shark scales!

## Do sharks have skin or scales?

# BOTH! The rough skin of a shark is covered in tiny teeth-like scales. People once used dried sharkskin as sandpaper.

# Did sharks and dinosaurs do battle?

Sharks have been around for 400 million years. They were alive during the time of Spinosaurus. This huge dinosaur could swim and hunt in water. It ate sharks for supper. Sharks snacked on dinosaurs, too. But they only feasted on the dead ones that washed out to sea. Dinner delivered!

## What's the biggest shark?

Some sharks are smaller than this book. Others are as long as a bus. The whale shark is the world's biggest shark. What does this mega shark eat? It eats mini-seafood called **plankton**!

Great white sharks are 15 to 20 feet (5 to 6 meters) long. That's about half the size of whale sharks. They take big bites out of fish, seals, and other sharks. *Chomp*!

A whale shark scoops up plankton with its wide mouth.

Spotted Dogfish shark.

## Are dogfish and catsharks related?

**YES!** Like wild dogs, dogfish sharks hunt in packs. These smallish sharks chase down fish and squid. Catsharks are cousins of dogfish. They have catlike eyes, and some are spotted or striped.

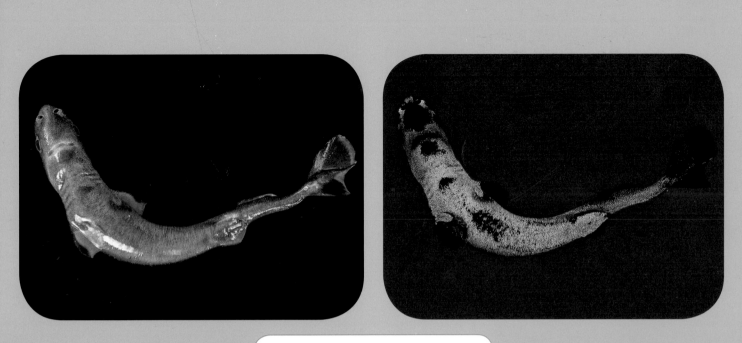

The glow of the pygmy shark helps hide it from predators.

## Do sharks glow in the dark?

Many ocean creatures make light, including small sharks. The glowing bellies of pygmy sharks help hide them from hungry hunters below. How? When predators look up at the pygmy shark, its glow blends in with sunlit water overhead.

One kind of lanternshark has glowing spines that look like light sabers.

# Who takes care of the babies?

Shark babies are called "pups." Sharks are born in different ways. But all shark pups take care of themselves. Sharks don't do the parenting thing.

Hammerhead shark pups swim together.

A mermaid's purse holds baby sharks.

Catsharks hatch from big egg cases called "mermaid purses." Tiger sharks, nurse sharks, great whites, and other large sharks are born alive in groups of six to eight pups. Happy Birthday!

## Can sharks fly?

A shark's body is shaped like an airplane. This shape helps the shark move easily through water. A shark's engine is its powerful swishing tail. It pushes the shark forward. A shark's side fins work like airplane wings. Water flowing around them lifts the shark off the seafloor or up toward prey.

A shark can "fly" through water!

Mako shark.

Fish avoid a young blacktip reef shark.

A gray reef shark swims while sleeping.

## When do sharks sleep?

While they're swimming! For most sharks, breathing only happens while moving. Swimming pushes air-filled water over their **gills** so they can breathe. If they stop swimming, they will die. Good thing sharks can swim and sleep at the same time!

Not all sharks have to nap while swimming. Nurse sharks have muscles that push water over their gills. They can breathe while staying still, just as most fish do.

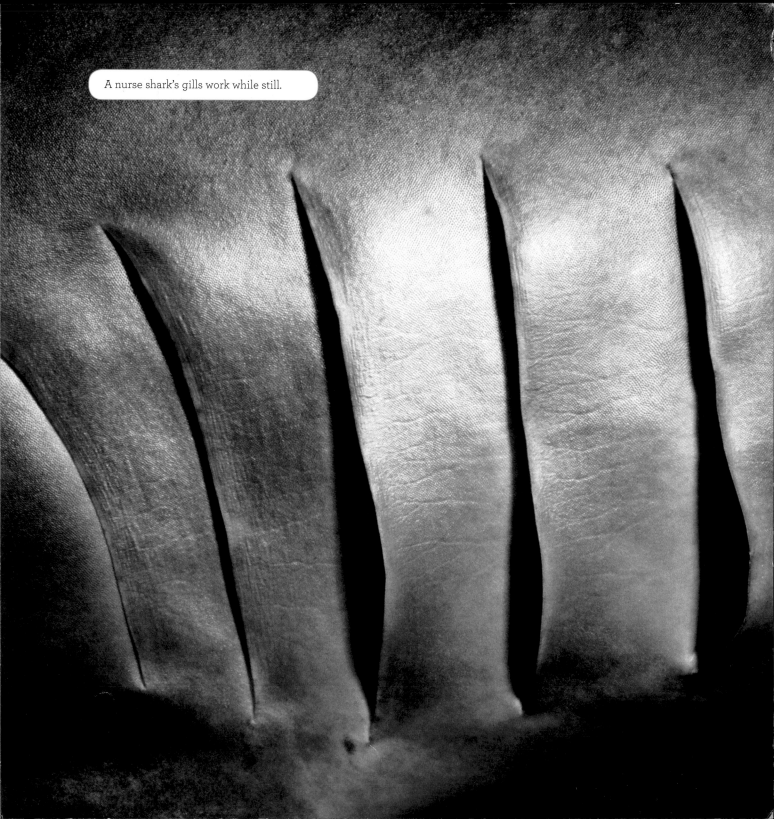

## *Do sharks have permanent teeth?*

# NOPE! Shark teeth don't last long. They get used up, fall out, and are replaced . . . over and over again! Shark mouths are filled with rows of teeth that are like conveyor belts. Shiny, sharp, new ones are always replacing dull old ones.

A shark might go through thousands of teeth during its lifetime. Some sharks use up a whole set of teeth almost every week!

# Can sharks sense electricity?

Yes! Like you, sharks have eyes that see, noses that smell, ears that hear, and mouths that taste. Sharks feel with an organ called a **lateral line**. It senses movement in the water. This organ runs along the sides of a shark's body.

Lateral line on a blacktip reef shark.

Sharks also have a sixth sense. Tiny **pores** on their heads pick up **electricity**. Sharks use it to sense weak electric signals given off by prey. Gotcha!

Hammerhead shark.

## What's the weirdest shark?

It's hard to decide! Goblin sharks have pop-out jaws.
These hidden jaws shoot out and snap prey by surprise.
*Whoosh*...snap!

The odd-shaped head of a hammerhead shark gives it
an all-around view. It can even see behind itself.

A sawshark looks like a living hedge trimmer. It smacks fish and squid with its long toothy snout.

Carpetsharks look like seaweed-covered rugs. These **camouflage** experts live on the ocean floor. They blend right in!

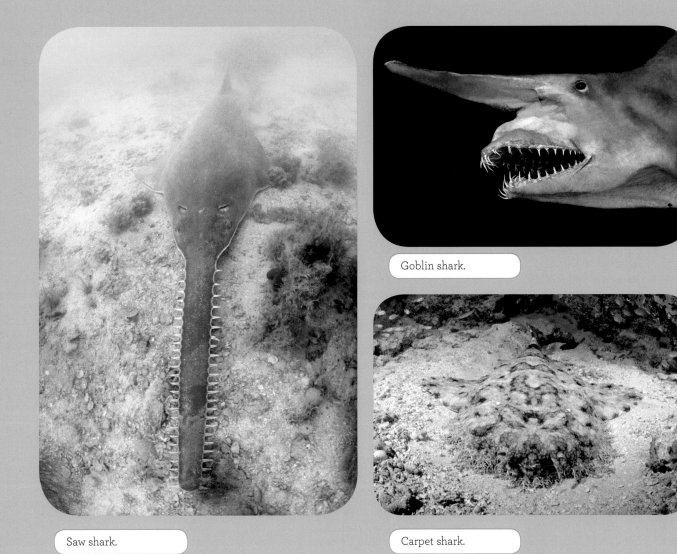

Goblin shark.

Saw shark.

Carpet shark.

## Do sharks eat people?

Sharks rarely attack people. Unlike yummy seals, surfers and swimmers are bony and bad-tasting to a shark. Sharks kill about six people a year in the whole world.

Pickled dogfish.

Sharkfin soup.

## Do people eat sharks?

Sharks in every ocean are fished by people. Dogfish sharks are often found in fish and chips. Soup made with shark fins is a pricey treat.

People kill about 100 million sharks a year. Some sharks are **endangered** because of too much fishing. Ocean pollution also harms sharks.

# Help the Sharks!

These groups are working to protect sharks.
Find out how:

- Shark Trust: www.sharktrust.org
- Shark Angels: www.sharkangels.org
- Project Aware: www.projectaware.org/sharks

# Shark Words to Know

**camouflage** – blending in with surroundings

**carnivore** – meat-eater

**cartilage** – strong, bendable material in bodies

**electricity** – a kind of energy

**endangered** – in danger of dying off

**gill** – a body part for breathing

**lateral line** – a sense organ

**plankton** – tiny animals and plants living in water

**pores** – small holes in skin

**predator** – an animal that hunts to eat

**prey** – an animal hunted for food

# Index